IT'S SCIENCE!

Hearing Sounds

Sally Hewitt

CHILDREN'S PRESS®

A Division of Grolier Publishing

NEW YORK • LONDON • HONG KONG • SYDNEY
DANBURY, CONNECTICUT

First published by Franklin Watts 1988

First American edition 1998 by
Children's Press
A Division of Grolier Publishing
90 Sherman Turnpike
Danbury, CT 06816
Visit Children's Press on the Internet at:

http://publishing.grolier.com

Series editor: Rachel Cooke
Art director: Robert Walster
Designer: Mo Choy
Picture research: Susan Mennell
Photography: Ray Moller unless otherwise acknowledged
Series consultant: Sally Nankivell-Aston

ISBN 0-516-20841-1
A CIP catalog record
for this book is available from
the Library of Congress.

Printed in Malaysia

Acknowledgments:
Bruce Coleman pp. 7tr (Gunter Ziesler), 15tr (Hans Reinhard);
Robert Harding p. 7br; John Walmsley p. 15l;
Performing Arts Library pp. 16t (Colin Willoughby), 21tl (Clive Barda);
Steve Shott p. 7l.
Thanks to our models: Jasmine Sharland, David Watts, Nakita Ogugua,
Harry Cudlipp, Shelley Tester and Ben Brooke.

Contents

Ringing Out

An alarm clock rings very loudly and wakes you up.
It may be the first sound a person hears in the morning.

Can you tell from the pictures what other sounds you might hear while you get ready for the day?

 TRY IT OUT!

Sit very still, shut your eyes, and listen for a few minutes. How many different sounds can you hear? Do you know what is making the sounds?

A bird singing, a friend talking on the telephone, and traffic driving along are all sounds you might hear every day.

THINK ABOUT IT!

Which sounds are nice to hear? Which sounds are unpleasant to hear? Can you see something on these two pages that makes a sound you don't like hearing?

7

Make a Noise!

The lump you can feel in the middle of your throat is called your **larynx**, or **voice box**. You can use your voice to make all kinds of different sounds.

TRY IT OUT!

Put your hand lightly over your voice box and feel what happens to it when you try doing these four things:

★ whispering
★ shouting
★ singing
★ humming

You can use your fingers, hands, and feet to make a noise, too.

TRY IT OUT!

Wag one finger, wave one hand, and shake one foot around. Did you make much noise? Now try clicking your fingers, clapping your hands, and stamping your feet. Listen to the noise you make now!
Can you tell why you made much more noise this time?

Keep Quiet!

Juice standing in a glass makes no sound. But pouring juice from a jug into a glass makes a noise. If you can hear a sound, it means that something is moving!

 **TRY IT OUT!**

Sit as still as you possibly can. Can you hear yourself making a noise? Stand up and dance about. Can you hear yourself making a noise now?

 THINK ABOUT IT!

Can you think of anything that is quite still that makes a sound?

10

This tissue paper, oaktag, wooden spoon, saucepan, and recorder are not making a noise. They are lying quite still on the table.

What could you do to make a noise with them?

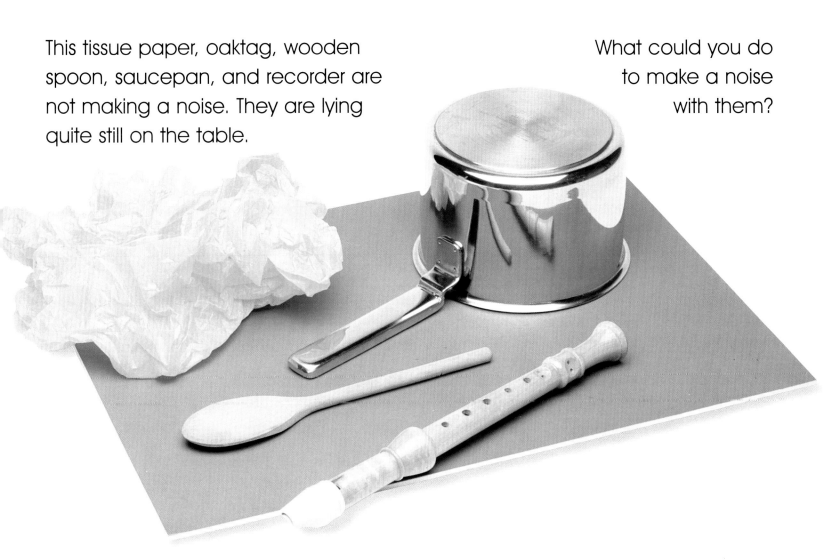

TRY IT OUT!

Collect some objects like these. Crumple the tissue paper, flap the oaktag, bang the saucepan with the spoon, and blow the recorder. Listen to the sounds they make now!

Listen . . .

The part of your ear that you can see is called the **outer ear**.

The outer ear is a very good shape for collecting sound.

TRY IT OUT!

Turn on a radio and turn the **volume** down so that it sounds quiet.

Cup your hand round one ear to make your outer ear bigger and point it towards the radio. Can you hear the radio more clearly?

Cover both ears with your hands. Can you hear the radio at all?

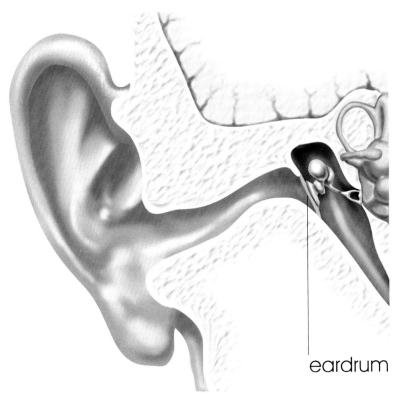

eardrum

Did you know you have a drum inside your ear called an **eardrum**?

When something makes a noise, the air all around it moves. The moving air hits your eardrum and makes it **vibrate**. Vibrate means to move back and forth very fast. When your eardrum vibrates, you hear the sound.

 TRY IT OUT!

Cover a bowl very tightly with cellophane. Sprinkle some rice on the cellophane. Bang a metal baking pan as loudly as you can just above the bowl. What happens?

The noise of the bang moves the air around it. The moving air vibrates the cellophane and makes the rice jump! The sound of the bang makes your eardrum vibrate, too.

13

Loud and Quiet

Some sounds are loud and some are quiet. The volume of a sound is how loud or quiet it is.

If you turn the volume up on the television or radio, you make the sound louder.

We can make loud and quiet sounds.

 **TRY IT OUT!**

Try singing, clapping, and tapping your foot quietly.
Now try singing, clapping, and stamping your foot very loudly. Did you work harder making the quiet sounds or making the loud sounds?

14

Some sounds, like a pin dropping onto the floor, are so quiet you can hardly hear them.

Other sounds are so loud that they could damage your eardrums.

This worker is wearing ear muffs to protect his ears from the sound of the jackhammer.

LOOK AGAIN

Look again at page 7. Put these sounds in order, too, starting with the quietest.

Have you heard the sounds made by all the things on this page? Can you put them in order, starting with the quietest?

15

High and Low

Sounds can be high or low as well as loud or quiet. The **pitch** of a sound is how high or low it is. Listen for high and low sounds like the ones on this page.

This big bass drum makes a low-pitched sound. Men's voices are low pitched, too.

The sound of this tin whistle is high-pitched, so are women's and children's voices.

 TRY IT OUT!

Put your hand lightly on your voice box. Sing a high **note** and gradually make your voice go as low as you can. Feel what happens to your voice box.

The thick strings on a guitar play low notes and the thin strings play higher notes.

 TRY IT OUT!

Find two glass bottles both the same size. Tap them gently with a metal spoon and listen to the sound they make.

Add a little water to one bottle and a larger amount to the other. Do you hear a high note or a low note when you tap them now?

Which bottle has the most air inside it?

THINK ABOUT IT!

How would you play the guitar loudly?
How would you play it quietly?

Making Music

You have to pluck, scrape, blow, bang, or rattle these musical instruments to play them.

Can you tell from the pictures what you would have to do to each one to make a sound?

A recorder is a **wind instrument**.

When you blow into it, you make the air inside
it vibrate and you hear a sound.

Which of the instruments on the opposite
page is a wind instrument?

 TRY IT OUT!

Collect some tubes and
empty plastic bottles like
these. Try blowing gently across
the top of them to make the
air inside them vibrate.
Can you hear a sound?

19

Twang and Bang!

A guitar and a cello are **stringed instruments**. They have big wooden **sound boxes** full of air.

You pluck the guitar strings with your fingers and scrape the cello strings with a bow to make a sound. When you do this, the air inside the sound boxes vibrates and makes a sound.

 TRY IT OUT!

Stretch an rubber band between your fingers and pluck it. What can you hear?

Stretch rubber bands around open containers to make a sound box. Pluck the elastic bands over the open part of the containers. What can you hear now?

A drum is a **percussion instrument**. Percussion instruments are banged or rattled to make a sound.

LOOK AGAIN

Look again at page 18 to find two more percussion instruments.

TRY IT OUT!

Collect pairs of things that you can bang together.

Fill containers with pasta, buttons, rice, or shells. Put the lids on tightly and rattle them.

Collect boxes, tins. and saucepans. Turn them upside down. Try banging and tapping them using different things for sticks.

How many different sounds can you make?

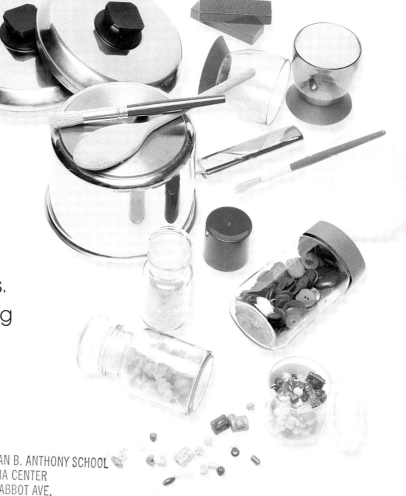

Traveling Sound

Sound from the radio travels through the air to your ears and you can hear it. Sound travels through bricks and wood as well as air.

 TRY IT OUT!

Turn on a radio and stand near it. You will be able to hear the sound clearly. Now go out of the room and shut the door. Does the sound of the radio travel through the door?

Put one ear to the wall and block the other ear. Can you hear the radio now?

Have you ever tried calling "hello!" in a big empty hall, or in a subway? If you have, you probably heard "hello!" immediately after, as if someone was calling back to you.

The sound of your voice travels through the air, hits the walls, and bounces back to you. We call this an **echo**.

Sound travels along string and wires.

 TRY IT OUT!

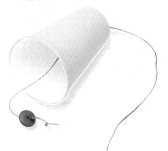

Make a hole in the bottom of two plastic cups. Thread a long piece of string through the holes and fix it with two buttons. Keep the string tight and talk to a friend through the telephone.

Sending Messages

You don't have to be very far away from someone before they find it difficult to hear what you are saying.

 TRY IT OUT!

Go somewhere with plenty of space, like a park or a playground. Say something to a friend so that they can hear you easily.
Now move farther away and say it again.
Keep moving farther away until you are too far away for your friend to hear what you are saying.

 **THINK ABOUT IT!**

How could you use flags, pen and paper, and a flashlight to send messages to someone who is too far away to hear you?

24

When you talk on the telephone, your voice can be sent along wires from house to house and even through the air from country to country.

You can see pictures and hear sounds on television from all around the world.

Radio stations **broadcast** sounds through the air for us to listen to on our radios.

 LOOK AGAIN

Look again at page 12. How do you hear messages from the radio?

25

Recording sound

Music, words, and other sounds can be recorded onto records, CDs, and tapes.

When they are played back, you hear the sound through **loudspeakers**. You can listen to them again and again.

Do you have any records, CDs, or tapes at home?
What do you like listening to?

 THINK ABOUT IT!

How could you listen to your favorite music if there were no records, CDs, or tapes to listen to?

You can make a **recording** of your own if you have a **microphone**, a tape recorder, and a tape.

![TRY IT OUT icon] **TRY IT OUT!**

You can read a story, sing, or play an instrument, tell a joke, or record a message to send to someone who lives far away. Listen to the recording you have made.

Useful Words

Broadcast Television and radio stations send out or broadcast pictures and sounds as invisible radio signals. Radio and television sets pick up these signals and change them back into sound and pictures.

Eardrum You have an eardrum inside your ear. It is a bit like the part of a drum you hit. When soundwaves hit your eardrum they make it vibrate and you hear that sound.

Echo When sound bounces off something, such as a wall, you hear the sound again. This is called an echo.

Larynx You can feel your larynx or voice box in your neck. When you speak, special cords in your larynx vibrate to make a sound.

Loudspeakers You hear the sound from televisions, radios, and CD players through loudspeakers. They amplify sound.

Microphone A microphone picks up sounds and turns them into electric signals. When you make a recording, you need a microphone to pick up the sound of your voice.

Note When you listen to music, each individual sound you hear is a note. You can sing notes or play them on musical instruments.

Outer ear Your outer ear is the part of your ear that you can see. It is a good shape for collecting sounds.

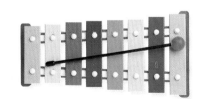

Percussion instruments Percussion instruments are banged, rattled, or tapped to make a sound. Drums, tambourines, and maracas are types of percussion instrument.

Pitch The pitch of a sound is how high or low it is. Each sound or note in music has a different pitch. Very high pitched sounds can be squeaky. Low pitched sounds are deep and growly.

Recording A recording is a copy of sound or music you can listen to again and again. You can make a recording of all kinds of sounds with a microphone and a tape recorder.

Sound box Some instruments have a big wooden sound box full of air. When the instrument is played, the air inside the sound box vibrates and makes the sound louder.

Stringed instruments Stringed instruments have strings stretched tightly across a sound box. They are played by plucking the strings or scraping them with a bow. Violins, cellos and guitars are all types of stringed instruments.

Vibrate To move a small distance very quickly back and forth.

Voice box See **larynx**.

Volume The volume of a sound is how loud or quiet it is. When we whisper the volume is quiet, and when we shout the volume is loud.

Wind instruments Wind instruments are played by blowing into them. The air inside the instrument vibrates and makes a sound. Flutes, recorders, and trumpets are types of wind instruments.

Index

About this book

Children are natural scientists. They learn by touching and feeling, noticing, asking questions, and trying things out for themselves. The books in the *It's Science!* series are designed for the way children learn. Familiar objects are used as starting points for further learning. *Hearing sounds* starts with a ringing alarm and explores the concept of sound.

Each double page spread introduces a new topic, such as volume and pitch. Information is given, questions are asked and activities are suggested that encourage children to make discoveries and develop new ideas for themselves. Look out for these panels throughout the book:

TRY IT OUT! indicates a simple activity, using safe materials, that proves or explores a point.
THINK ABOUT IT! indicates a question inspired by the information on the page but which points the reader to areas not covered by the book.
LOOK AGAIN introduces a cross-referencing activity which links themes and facts through the book.

Encourage children not to take the familiar world for granted. Point things out, ask questions, and enjoy making scientific discoveries together.